SCOTTISH T

Recipes from Scotland

by Sue Mc Dougall

Sphere Books Ltd., 30-32 Gray's Inn Road, London WC1X 8JL

First published 1976
First published in this edition 1983
Reprinted 1984

Use plain flour unless otherwise stated, castor sugar is recommended for butter and sugar creamed mixtures.

Printed and bound in Great Britain by BKT

PARKIN

8 oz. (200 g.) flour
4 oz. (100 g.) butter
8 oz. (200 g.) medium oatmeal
4 oz. (100 g.) caster sugar
6 oz. (150 g.) golden syrup
6 oz. (150 g.) black treacle
2 teaspoons (10 ml.) baking powder
2 teaspoons (10 ml.) ground ginger
1 egg
4 tablespoons (60 ml.) milk
Blanched almonds

Sift the flour, baking powder and ginger together. Rub the butter into the flour mixture and stir in the oatmeal and sugar. Warm the syrup and treacle and pour into a well in the middle of the flour/oatmeal mixture. Lightly beat the egg and add with the milk. Mix well. Turn, into a lined and greased 9 inch (23 cm.) square tin. Place the blanched almonds on top. Bake in a moderate oven (350°F, 177°C, gas mark 4) for 1 hour. When cool turn out and cut into squares.

SHORTBREAD

8 oz. (200 g.) flour
4 oz. (100 g.) butter
2 oz. (50 g.) caster sugar

Mix the flour and sugar and rub in the butter. Knead well. Roll out and cut into strips. Prick all over. Bake in a moderate oven (350°F, 177°C, gas mark 4) for 30 – 40 minutes until lightly browned. Cut into biscuits before cold.

GINGER SHORTBREAD

Add ½ teaspoon (3 ml.) ground ginger to the flour in the recipe for shortbread.

DUNDEE CAKE

8 oz. (200 g.) flour
6 oz. (150 g.) butter
6 oz. (150 g.) brown sugar
2 oz. (50 g.) chopped mixed peel
3 oz. (75 g.) currants
3 oz. (75 g.) sultanas
3 oz. (75 g.) stoned raisins
2 oz. (50 g.) blanched almonds
1 teaspoon (5 ml.) baking powder
Grated rind of 1 lemon
4 eggs

Cream the butter and sugar until light and fluffy. Add the lemon rind. Beat the eggs and beat into the butter mixture. Sieve the flour and baking powder together. Fold the flour, fruit, peel and half the blanched almonds, chopped, into the mixture. Turn into a greased and lined 8 inch (20 cm.) cake tin. Arrange the almonds left on top of the cake. Bake in a moderate oven (350°F, 177°C, gas mark 4) for 2½–3 hours. If the top browns too quickly, cover with brown paper or greased paper. Turn out and cool on a wire rack.

SCOTTISH BLACK BUN

Shortcrust pastry
Filling:
8 oz. (200 g.) plain flour
4 oz. (100 g.) brown sugar
1 lb. (400 g.) seedless raisins
1 lb. (400 g.) currants
2 oz. (50 g.) chopped mixed peel
4 oz. (100 g.) chopped blanched almonds
1 teaspoon (5 ml.) ground cinnamon
1 teaspooon (5 ml.) bicarbonate of soda
1 teaspoon (5 ml.) cream of tartar
¼ pint (125 ml.) whisky
4 tablespoons (60 ml.) milk
1 egg

Make the shortcrust pastry as on page 9. Grease an 8 inch (20 cm.) cake tin. Roll out two-thirds of the pastry and line the tin so that the pastry comes above the edge of the tin.

Filling: Sift the flour, spices, cream of tartar and bicarbonate together. Stir in the sugar, fruit, chopped peel, almonds and mix well. Lightly beat the egg and mix with the milk and whisky. Pour into the fruit mixture and mix well. Turn into the pastry case. Turn the edges of the pastry inwards over the top of the filling. Roll out the remaining pastry. Moisten the edges of the pastry and put on the lid. Seal the edges firmly. Prick all over with a fork and glaze with beaten egg. Make several holes in the lid using a skewer and passing the skewer to the bottom of the cake. Bake in a moderate oven (350°F, 177°C, gas mark 4) for 2½ –3 hours. If the cake becomes too brown, cover with greaseproof paper. Turn on to a wire rack to cool. Store for several weeks before cutting.

SCOTTISH GINGER CAKE

12 oz. (300 g.) plain flour
6 oz. (150 g.) butter
3 oz. (75 g.) brown sugar
12 oz. (300 g.) black treacle.
2 oz. (50 g.) sultanas
4 oz. (100 g.) chopped peel
2 oz. (50 g.) chopped preserved ginger
½ teaspoon (3 ml.) salt
2 teaspoons (10 ml.) bicarbonate of soda
1 teaspoon (5 ml.) ground ginger
3 eggs
3 tablespoons (45 ml.) milk

Sift the flour, salt, bicarbonate of soda and ginger together. Stir in the sultanas, chopped peel and preserved ginger. Warm the treacle, butter and sugar. Lightly beat the eggs with the milk. Add to the treacle mixture. Pour into a well in the middle of the flour/fruit mixture and beat thoroughly. Turn into a lined 7 inch (18 cm.) square cake tin and bake in a moderate oven (325°F, 163°C, gas mark 3) for 1¼ hours. Take out of the tin and place on a wire rack to cool. Store in a tin.

RASPBERRY TART

Shortcrust Pastry:
8 oz. (200 g.) plain flour
4 oz. (100 g.) butter
3 tablespoons (45 ml.) cold water

Filling:
Raspberries
Caster sugar

Rub the butter into the flour until the mixture looks like breadcrumbs. Add the water and form into a ball. Roll out two thirds of the pastry and line a deep plate. Wash and hull the raspberries and place in the tart. Sprinkle with sugar to taste. Roll out the rest of the pastry to make the lid. Brush with milk and bake in a fairly hot oven (400°F, 204°C, gas mark 6) for 25 minutes until the pastry is cooked. To serve, dust with caster sugar and top with whipped cream.

SCONES

8 oz. (200 g.) self-raising flour
2 oz. (50 g.) butter
1 teaspoon (5 ml.) baking powder
½ teaspoon (3 ml.) salt
¼ pint (125 ml.) milk

Sift the dry ingredients together and rub in the butter until the mixture looks like breadcrumbs. Stir in enough milk to form a soft dough. Turn on to a floured board and knead lightly. Roll out to a thickness of ¾ inches (2 cm.) and cut into 2-inch (5 cm.) rounds. Bake on a greased baking sheet in a very hot oven (450°F, 232°C, gas mark 8) for 10 minutes. Serve split and buttered.

FRUIT SCONES

Stir 2 oz. (50 g.) sultanas, currants or mixed fruit into the dry ingredients in the scone mixture.

GIRDLE SCONES

Make the scone mixture and cook on a fairly hot girdle, allowing 3 – 4 minutes a side.

CHEESE SCONES

8 oz. (200 g.) self-raising flour
1½ oz. (40 g.) butter
4 oz. (100 g.) grated cheese
1 teaspoon (5 ml.) baking powder
1 teaspoon (5 ml.) mustard
Pinch of salt
½ pint (250 ml.) milk

Sift the flour, baking powder and salt together. Rub in the butter until the mixture looks like breadcrumbs. Stir in half the cheese and the mustard. Add enough milk to give a fairly soft dough. Roll out on a floured board to a thickness of ¾ inch (2 cm.) and cut into 2-inch (5 cm.) rounds. Sprinkle with the rest of the cheese. Bake in a hot oven (425°F, 218°C, gas mark 7) for 10 minutes. Serve split and buttered.

BUTTERFLY CAKES

6 oz. (150 g.) self raising flour
4 oz. (100 g.) butter
4 oz. (100 g.) caster sugar
2 eggs
1 teaspoon (5 ml.) grated lemon rina
1 tablespoon (15 ml.) milk

Topping:
¼ pint (125 ml.) whipping cream
1 teaspoon (5 ml.) caster sugar
Few drops vanilla essence
Icing sugar

Cream the butter and sugar until the mixture is light and fluffy. Lightly beat the eggs and add one at a time. Sieve the flour. Fold the flour and grated lemon rind into the mixture. Add the milk to give a soft, dropping consistency. Bake in a moderately hot oven (375°F, 191°C, gas mark 5) for 15 minutes until the cakes are golden brown and firm to the touch. When cool, cut the top off each cake. Whip the cream and add the caster sugar and vanilla essence. Place a teaspoonful of cream on each cake and replace the cut tops. Dredge with icing sugar.

SANDWICHES

Use brown and white bread, and rolls. Soften the butter to make spreading easier. Fillings should be tasty. Wrap sandwiches in foil or polythene and keep in a refrigerator. Avoid over-moist fillings which make the bread soggy.

SWEET FILLINGS

1. *Jam*
2. *Chopped dates and chopped sweet apple*
3. *Honey and banana*
4. *Lemon curd and chopped sweet apple*
5. *Chopped nuts and honey*

SAVOURY FILLINGS

1. *Minced cooked beef and horseradish sauce*
2. *Ham, chopped chives (or onions) and mustard*
3. *Tongue and chopped tomato*
4. *Corned beef, lettuce and mayonnaise*
5. *Lettuce and tomato*
6. *Tongue and cucumber*
7. *Grated cheese and onion*
8. *Mashed egg and chopped chives*
9. *Meat pâté and sliced cucumber*
10. *Chopped cooked bacon and mushrooms*

Mix ingredients together, blending with a little butter or mayonnaise if necessary. Season to taste.

CHEESE, APPLES AND NUTS

½ oz. (15 g.) butter
1 oz. (25 g.) grated cheese
1 sweet chopped apple
2 teaspoons (10 ml.) chopped nuts

Mix all the ingredients together.

CHICKEN AND HAM

½ oz. (15 g.) butter
1 oz. (25 g.) minced chicken
1 oz. (25 g.) minced ham
1 tablespoon (15 ml.) chopped chives
1 tablespoon (15 ml.) mayonnaise
Salt and pepper

Mix all the ingredients together. Season.

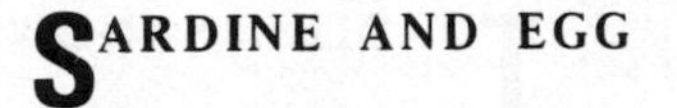

SARDINE AND EGG

1 tin sardines
1 egg
½ teaspoon (3 ml.) Worcestershire sauce
Salt and pepper

Hard boil the egg. Remove the heads, tails and backbones of the sardines. Mix all the ingredients together. Season.

EGG AND TOMATO

½ oz. (15 g.) butter
1 egg
1 tomato
Salt and pepper

Hard boil the egg. Place tomato in boiling water to remove skin. Remove the seeds and cut up. Mix egg, tomato and butter together. Season.

TOASTED SANDWICHES

2 pieces buttered toast

***Fillings*:**

1. *Fried egg and bacon*
2. *Baked beans and bacon pieces*
3. *Chopped mushrooms, bacon pieces and mustard*
4. *Flaked kipper, tomato and mustard*
5. *Chopped ham and mustard*

STUFFED FRENCH LOAF

1 French loaf
1 oz. (25 g.) butter
Filling:
Lettuce leaves
Boiled or scrambled egg
Sliced cucumber
Chopped tomato
Chopped onions or chives
Mayonnaise

Cut the loaf in half lengthwise and remove most of the crumb to leave two crusty shells. Butter and fill with salad. Put the two halves together and cut in slices to serve.

POTTED BEEF

1 lb. (400 g.) stewing steak
¼ pint (125 ml.) stock **or** *water*
2 oz. (50 g.) butter
1 clove
1 blade mace
Salt and pepper

Remove the skin and as much fat as possible from the meat. Cut the meat into cubes. Place in an ovenproof dish with the stock **or** water, clove and mace. Season. Cook for 1½ - 2 hours in a moderate oven (350°F, 180°C, gas mark 4). Drain the liquid, remove the clove and finely mince the meat (or blend the meat in an electric blender). Melt the butter and mix half into the beef. Spoon into sterilized jars and seal with the rest of the melted butter. Keep in a refrigerator and use within 2 days. Serve on toast or brown bread.

POTTED SHRIMPS

¼ lb. (100 g.) shrimps (cooked or frozen)
2 oz. (50 g.) butter
Pinch ground mace
Pinch ground nutmeg
Salt and pepper

Melt half the butter in a pan and add the shrimps. Season. Heat for 2 - 3 minutes. Sieve or gently mash. Spoon into sterilized jars. Melt the rest of the butter and pour over the tops of the shrimps.

HADDOCK ON TOAST

1 medium Findon fillet
¼ pint (125 ml.) milk
½ oz. (15 g.) flour
1 oz. (25 g.) butter
Salt and pepper
Toast

Wash the fish and simmer in the milk and butter for 5 - 7 minutes. Remove the fish, skin and flake the flesh. Strain the milk. Blend the flour with a little of the milk to form a smooth paste. Add to the rest of the milk and bring to the boil, stirring to keep the mixture smooth. Season. Boil for 2 - 3 minutes for the sauce to thicken. Add the flaked fish. Serve on buttered toast.

HADDOCK RAREBIT

1 Findon haddock
1 oz. (25 g.) flour
2 oz. (50 g.) cheese
½ pint (250 ml.) milk
2 eggs
2 tomatoes
Salt and pepper

Clean the fish and simmer gently for 7 minutes in salted water. Remove and flake the fish. Place the tomatoes in boiling water and skin. Slice the tomatoes and place with the fish in an ovenproof dish. Blend the flour with a little of the milk and then stir in the rest of the milk. Bring to the boil, stirring to keep the mixture smooth. Boil for 3 minutes. Grate the cheese, beat the eggs and add to the sauce. Heat until sauce thickens but don't boil. Pour over the fish and tomato.

SAUSAGE ROLLS

Shortcrust pastry
8 oz. (200 g.) sausage/meat
Flour
Milk to glaze

Make the shortcrust pastry as on page 9. Roll the pastry into an oblong and divide in two lengthwise. Divide the sausage meat in two and form into 2 rolls each the length of the pastry. Dust with flour and place on the pastry. Roll each strip over to enclose the sausage, closing the seam by flaking or scalloping. Brush with milk and cut into pieces 2 inches (5 cm.) long. Place on a baking sheet and bake in a fairly hot oven (400°F, 204°C, gas mark 6) for 15 minutes and then at 350°F, 177°C, gas mark 4 for a further 15 minutes to cook the sausage meat.

GLAZED BAKED HAM

5 lb. (2 kg.) gammon
3 tablespoons (45 ml.) clear honey
1 tablespoon (15 ml.) grated orange peel

Wash the ham and soak for 2 hours. Dry and wrap in aluminium cooking foil. Bake in a fairly hot oven (375°F, 191°C, gas mark 5) for 3 - 3½ hours. 30 minutes before the ham is cooked, score the fat of the ham into diamonds. Mix the honey and orange peel and coat the ham. Return to the oven and complete baking. Serve with salad.

RAISED VEAL AND HAM PIE

Hot water pastry
4 oz. (100 g.) ham
1 lb. (400 g.) shoulder of veal
2 hard boiled eggs
4 tablespoons (60 ml.) water or stock
½ pint (250 ml.) chicken stock
2 teaspoons (10 ml.) gelatine
Pinch mixed herbs
Pinch ground mace
Salt and pepper

Roll out two thirds of the pastry on a floured board, keeping the remainder warm and covered until ready for use. Lift on to the outside of a pie dish and mould into a pie-shape. Carefully remove the pie dish and place the pastry shell on to a greased baking sheet. Cut the meat into small pieces and mix with the herbs and seasoning. Fill the pastry shell with the meat mixture, placing the eggs in the middle. Wet the edges of the pie. Roll out the rest of the pastry to make a lid. Seal the edges and brush the lid with beaten egg or milk. Tie a piece of greaseproof paper around the pie. Cook in a fairly hot oven (425°F, 218°C, gas mark 7) for 20 minutes, and then at 350°F, 177°C, gas mark 4 for 2 hours until the meat is

cooked and feels tender when tested with a skewer. Dissolve the gelatine in the chicken stock. When the pie is cold, make two small holes in the lid and pour in the jelly stock. Leave to set.
To make 4 small pies: Use jam jars as pie moulds and cook for a total of 1 hour or until the meat is tender.

HOT WATER CRUST PASTRY

12 oz. (300 g.) flour
4 oz. (100 g.) lard or cooking fat
¼ pint (125 ml.) water or milk and water
1½ teaspoons (8 ml.) salt

Sift the flour and salt together. Melt the fat in the liquid and bring to the boil. Pour into a well in the flour mixture and stir. Work into a lump using a wooden spoon. Knead lightly until smooth. Place in a plastic bag or cover with a damp cloth and stand in a warm place for 30 minutes.

DUMFRIES PIE

Shortcrust pastry
3 oz. (75 g.) bacon
2 eggs
1 tomato
Salt and pepper

Make shortcrust pastry as on page 9. Using half the pastry line a flan tin or deep pie plate. Cut up the bacon. Place the tomato in boiling water. Skin and cut up. Mix with the bacon pieces and spread over the flan case. Lightly beat the eggs. Season and pour over the bacon mixture. Cover with the rest of the pastry. Brush with a little milk. Bake in a fairly hot oven (400°F, 204°C, gas mark 6) for 20 - 25 minutes.

CHEESE AND HAM FLAN

Shortcrust pastry
2 oz. (50 g.) cheese
2 oz. (50 g.) cooked ham
2 teaspoons (10 ml.) cornflour
½ pint (250 ml.) milk
1 egg
Salt and pepper

Make shortcrust pastry as on page 9. Line a flan tin or deep pie plate with the pastry. Coat with a little egg white and bake for 10 minutes in a fairly hot oven (400°F, 204°C, gas mark 6). Lightly beat the egg and blend with the cornflour. Add the milk. Grate the cheese and stir into the mixture, keeping a tablespoon (15 ml.) back. Dice the ham and stir into the mixture. Season. Pour into the flan case and sprinkle the remaining cheese on top. Bake in a moderate oven (350°F, 177°C, gas mark 4) for 20 minutes until the flan is set.

OATCAKES

½ lb. (200 g.) oatmeal
½ oz. (15 g.) bacon fat
½ teaspoon (3 ml.) baking powder
Pinch salt
Hot water

Mix dry ingredients together. Melt the fat and rub into the oatmeal mixture. Add enough hot water to bind the mixture, giving a rather soft consistency. Knead well and roll out into a round on an oatmealed board. Cut into four and bake on a hot girdle. When one side is cooked place in a cool oven to dry out.

OATMEAL BANNOCKS

3 oz. (75 g.) fine oatmeal
1 oz. (25 g.) self raising flour
1 oz. (25 g.) butter
Pinch bicarbonate of soda
Pinch of salt
Warm water

Mix the dry ingredients together and rub in the butter. Bind together with enough warm water to give a stiff consistency. Turn on to a board dusted with oatmeal and work into a round. Roll out to a thickness of ¼ inch (·5 cm). Keep the dough well dusted to prevent it from sticking. Form into a round using a plate. Sprinkle with oatmeal and cook on a warmed girdle until the edges begin to curl. Turn and cook the other side. The bannock may be cut into quarters or farls. Dough for more than one bannock cannot be prepared at the same time because it stiffens quickly.

SCOTCH TOAST

4 oatcakes
Bloater paste
2 eggs
1 oz. (25 g.) butter
Pinch of cayenne pepper
1 oz. (25 g.) browned breadcrumbs

Oatcakes should be hot. Melt the butter and add 2 egg yolks and 1 egg white, the bloater paste and pepper. Stir, until thick. Pile the mixture on top of the oatcakes and sprinkle with browned breadcrumbs.

SCOTCH RAREBIT

8 oz. (200 g.) Cheddar cheese
½ oz. (15 g.) butter
2 teaspoons (10 ml.) flour
2 teaspoons (10 ml.) Worcestershire sauce
3 tablespoons (45 ml.) whisky
½ teaspoon (2 ml.) dry mustard
Pepper
4 rounds of bread

Toast the bread on one side. Grate the cheese into a saucepan. Add the butter, Worcestershire sauce, flour and mustard and a shake of pepper. Mix well and add the whisky. Heat slowly until the cheese and butter have melted. When the mixture leaves the side of the pan, spread on to the untoasted sides of the bread. Place under a grill and brown.

MARROW CHEESE

1 marrow
2 oz. (50 g.) flour
2 oz. (50 g.) butter
6 oz. (250 g.) cheese
1 pint (500 ml.) milk
Salt
Few sprigs of parsley

Peel the marrow and remove the seeds. Cut into pieces and boil in salted water until the marrow is soft. Drain and place the marrow in an ovenproof dish. Melt the butter and stir in the flour to make a smooth paste. Slowly add the milk and bring to the boil, stirring all the time to avoid lumps. Boil for 3 minutes. Grate the cheese and stir in 4 oz. (100 g.) slowly. Pour the cheese sauce over the marrow. Sprinkle the remainder of the cheese on top. Bake in a moderate oven (350°F, 177°C, gas mark 4) for 20 minutes until the cheese has browned. Garnish with parsley.

CRUMPETS

4 oz. (100 g.) self-raising flour
½ oz. (15 g.) butter
½ oz. (15 g.) caster sugar
¼ pint (125 ml.) milk
2 eggs

Mix the flour and sugar together. Rub in the butter. Lightly beat the egg. Mix the egg and milk with the flour mixture, keeping the batter smooth. Drop spoonfuls on to a hot girdle. When brown on one side, turn and cook on the other side. Cool in a towel.

CHICKEN PANCAKES AND WHISKY SAUCE

Pancakes:

4 oz. (100 g.) plain flour
½ pint (250 ml.) milk
1 egg
Pinch of salt
Fat for frying

Sift the flour and salt together. Add the egg and a little milk to the flour. Mix well, avoiding the formation of lumps. Stir in the rest of the milk and beat until smooth. Cover and leave to stand for 1 hour. Coat a frying pan with fat and heat until hot. Pour a little of the batter into the frying pan, tilting the pan so that the base of the pan is thinly covered. When cooked on one side, turn over. Keep the pancakes warm.

Filling:

8 oz. (200 g.) cooked chicken
4 oz. (100 g.) mushrooms
½ oz. (15 g.) butter
1 oz. (25 g.) flour
1 pint (500 ml.) milk
3 tablespoons (45 ml.) whisky
2 teaspoons (10 ml.) chopped parsley
Salt and pepper

Melt the butter and work in the flour to make a smooth paste. Slowly add the milk, stirring to avoid the formation of lumps. Heat gently until the sauce thickens and cook for 3 minutes. Cut up the chicken and mushrooms and add to the sauce. Season and simmer until the mushrooms are cooked. Add the whisky and chopped parsley. Place a little in the centre of each pancake and roll up. Serve hot.

PLAIN OMELETTE

2 eggs
Knob of butter
1 tablespoon (15 ml.) cold water
Salt and pepper

Whisk the eggs but do not overbeat or make frothy. Season and add the cold water. Lightly grease the omelette pan with the butter and place over a gentle heat. When the pan is hot pour the egg mixture into the hot fat. Stir until the egg mixture sets. Cook for 1 more minute. Turn the top third of the omelette towards the centre and then turn the bottom third towards the centre. Turn on to a hot plate and serve at once.

CHEESE OMELETTE

Grate 1 oz. (25 g.) cheese. Place half of this in the centre of the omelette before folding. Sprinkle the rest of the cheese over the omelette after folding.

KIDNEY OMELETTE

Skin, core and chop 1 sheep's kidney. Add 1 teaspoon (5 ml.) chopped onion and fry in butter until tender. Place in the centre of the omelette before folding.

SHRIMP OMELETTE

Gently sauté 2 oz. (50 g.) shrimps in butter and place in the centre of the omelette before folding. Serve at once with a squeeze of lemon juice.

CHEESE CUSTARD

1 oz. (25 g.) cornflour
1 pint (½ litre) milk
4 oz. (100 g.) cheese
2 eggs
Salt and pepper
Mustard
Sprig parsley

Lightly beat the eggs. Blend the cornflour with the eggs. Stir in the milk, taking care to keep the mixture free from lumps. Add the cheese and season. Pour into a buttered fireproof dish and bake in a moderate oven (350°F, 177°C, gas mark 4) for 45 minutes.

CALEDONIAN CREAM

½ lb. (800 g.) cottage cheese
1½ tablespoons (25 ml.) marmalade
1 tablespoon (15 ml.) sugar
1 tablespoon (15 ml.) malt whisky
2 teaspoons (10 ml.) lemon juice

Mix all the ingredients together and beat. Freeze before serving.

SCOTCH TRIFLE

4 small trifle sponge cakes
6 macaroons
1 oz. (25 g.) ratafias
8 oz. (200 g.) can peaches
6 tablespoons (90 ml.) sherry **or**
6 tablespoons (90 ml.) brandy and water
½ pint (250 ml.) custard
½ pint (250 ml.) cream
½ teaspoon (3 ml.) vanilla essence
Strawberry jam
2 teaspoons (10 ml.) caster sugar
1 oz. (25 g.) almonds

Slice the sponges and spread with jam. Place at the bottom of a trifle bowl. Add the ratafias, macaroons and peaches (sliced), keeping a little fruit for decoration. Pour the sherry or brandy/water and vanilla essence over the mixture. When the custard is cool, pour into the bowl. Whip the cream and sugar and pile on top. Decorate with almonds and pieces of peach.

CUSTARD

½ pint (250 ml.) milk
2 eggs
1 oz. (25 g.) sugar
Vanilla essence

Lightly beat the eggs. Heat the milk and pour over the beaten eggs. Strain. Warm over a low heat, with constant stirring, until custard thickens. Do not boil. Add the sugar and stir.

BRANDY ROLLS

2 oz. (50 g.) flour
2 oz. (50 g.) caster sugar
2 oz. (50 g.) butter
2 tablespoons (30 ml.) golden syrup
1 teaspoon (5 ml.) ground ginger
1 teaspoon (5 ml.) grated lemon rind
1 teaspoon (5 ml.) brandy
Cream

Melt the butter, syrup and sugar. Stir in the other ingredients (except the cream) and mix well. Drop in small teaspoonfuls on a greased tin about 4 inches (10 cm.) apart. Bake in a moderate oven (350°F, 177°C, gas mark 4) for 8 – 10 minutes. Cool for 1 minute and then roll round the handle of a wooden spoon. Whip the cream and pipe into the ends of the brandy rolls.

PEACHES IN WHISKY

8 fresh peaches
½ lb. (800 g.) caster sugar
1 pint (500 ml.) water
3 tablespoons (45 ml.) malt whisky
Cream

Soak the peaches for 5 minutes in water. Skin. Dissolve the sugar in 1 pint (500 ml.) water and boil. Add the peaches and heat gently for 15 minutes. Remove the peaches. Add the whisky to the syrup and boil for 5 minutes. Cool and pour over the peaches. Chill. Serve with whipped cream.

ABERNETHY BISCUITS

8 oz. (200 g.) flour
4 oz. (100 g.) butter
1½ oz. (40 g.) caster sugar
1 teaspoon (5 ml.) cream of tartar
½ teaspoon (3 ml.) bicarbonate of soda
1 tablespoon (15 ml.) milk
Pinch of Salt

Sift the flour, salt, bicarbonate of soda and cream of tartar together. Rub in the butter until the mixture looks like breadcrumbs. Dissolve the sugar in the milk and stir into the flour/butter mixture. Form into a stiff dough. Roll out to a thickness of ¼ inch (½ cm.). Cut into squares and prick all over. Bake in a moderate oven (350°F, 177°C, gas mark 4) for 20 minutes.

Metric equivalents: The metric quantities given in brackets where necessary are not the exact equivalents of the imperial quantities. The metric quantities indicate suitable amounts. We are all used to doubling or halving a recipe. Thus a Victoria sandwich may be made using 4 oz. each of flour, sugar and butter with 2 eggs, or 6 oz. each of flour, sugar and butter with 3 eggs. 'Going metric' is just another variation of this. The proportions of the ingredients remain unchanged: in this case, 100 g. each of flour, sugar and butter with 2 eggs. Liquids should be added carefully, especially when eggs are also used, to obtain the correct consistency to allow for the fact that 1 oz = 28·35 g. and 1 pint = 568 ml.

1 American cup contains:-

Butter, 8 oz.
Cheddar cheese, 4 oz.
Breadcrumbs (dry) 4 oz.
Flour - plain, self-raising, 4 oz.

Currants, 5 oz.
Sugar - granulated, caster, 7 oz
icing, 4½ oz.

Index